# The Great Barrier Reef

## Using Graphs and Charts to Solve Word Problems

Therese Shea

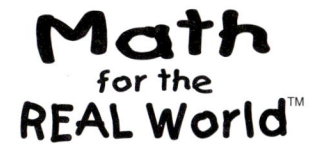

**Math for the REAL World**™

Rosen Classroom Books & Materials™
New York

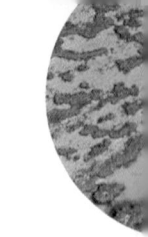

Published in 2006 by The Rosen Publishing Group, Inc.
29 East 21st Street, New York, NY 10010

Copyright © 2006 by The Rosen Publishing Group, Inc.

All rights reserved. No part of this book may be reproduced in any form without permission in writing from the publisher, except by a reviewer.

Book Design: Haley Wilson

Photo Credits: Cover © DigitalVision; pp. 3–32 (border) © Yann Arthus-Bertrand/Corbis; p. 4 Digital Image © 1996 Corbis; original image courtesy of NASA/Corbis; pp. 7, 21 © Jeffrey L. Rotman/Corbis; p. 8 © Gary Bell/Australian Picture Library/Corbis; p. 11 © Visuals Unlimited/Getty Images; p. 13 © Ralph A. Clevenger/Corbis; p. 15 © Taxi/Getty Images; p. 17 © Kevin Schafer/Corbis; p. 19 © Lawson Wood/Corbis; p. 25 © Stephen Frink/Corbis.

Library of Congress Cataloging-in-Publication Data

Shea, Therese.
   The Great Barrier Reef : using graphs and charts to solve word problems / Therese Shea.
     p. cm. — (Math for the real world)
   Includes index.
   ISBN 1-4042-3359-8 (lib. bdg.)
   ISBN 1-4042-6071-4 (pbk.)
   6-pack ISBN 1-4042-6072-2
   1. Problem solving—Graphic methods—Juvenile literature. 2. Word problems (Mathematics)—Juvenile literature. 3. Great Barrier Reef (Qld)—Juvenile literature.   I. Title. II. Series.
   QA63.S443 2006
   510—dc22
                                  2005013891

Manufactured in the United States of America

CPSIA Compliance Information: Batch #WR412180RC: For further information contact Rosen Publishing, New York, New York at 1-800-237-9932.

# Contents

| | |
|---|---|
| A World Wonder | 5 |
| What Is the Reef Made Of? | 6 |
| Creatures of the Reef | 12 |
| Endangered Species | 16 |
| Enemies of the Great Barrier Reef | 20 |
| Shipwreck! | 24 |
| Touring the Reef | 28 |
| Valued by the World | 30 |
| Glossary | 31 |
| Index | 32 |

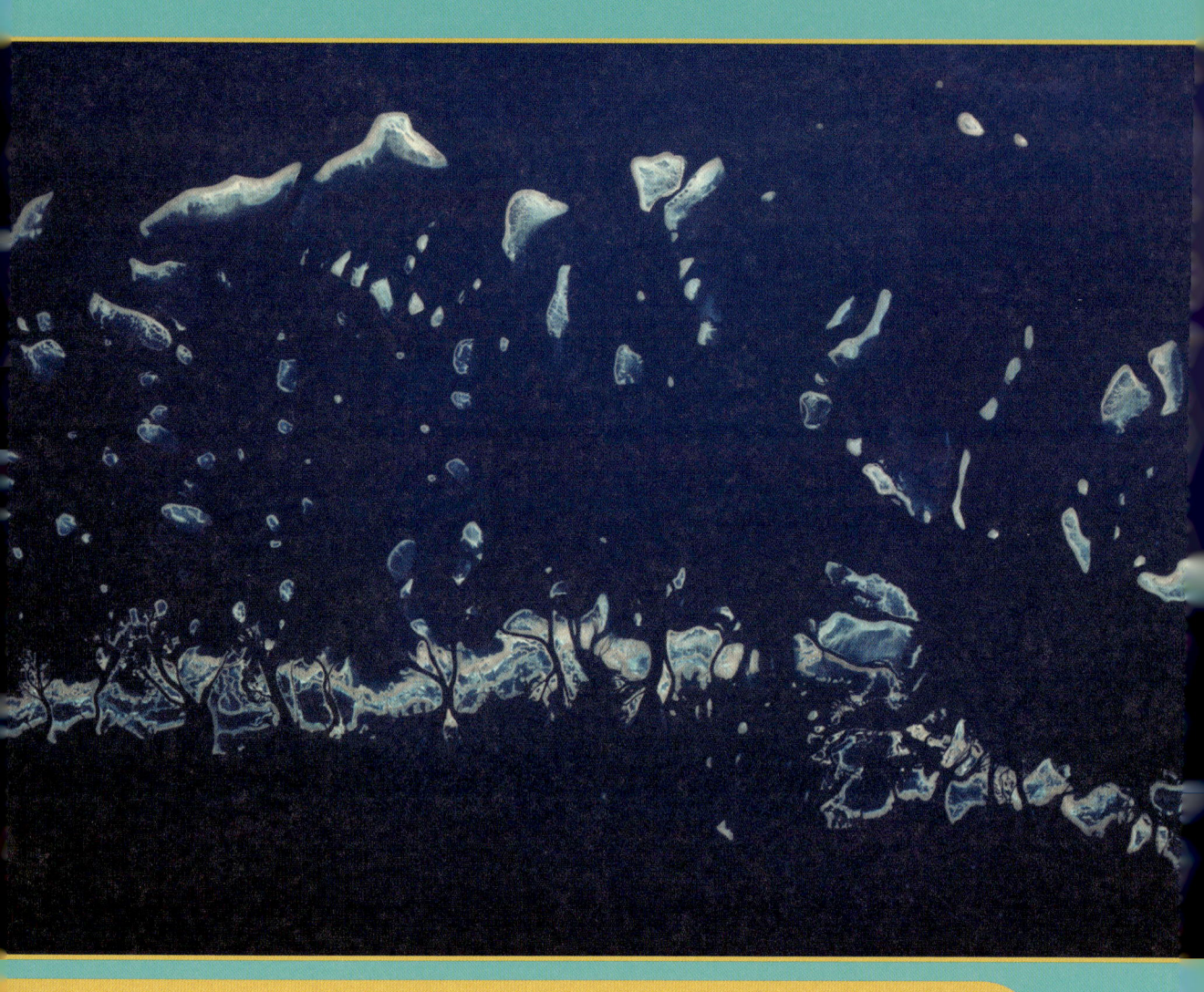

This photograph of the Great Barrier Reef was taken from outer space.

# A World Wonder

A wonder is something that fills you with awe and amazement. There are seven places on Earth known as the Seven Wonders of the Natural World. One of these seven wonders is the Great Barrier Reef.

The Great Barrier Reef is the world's largest coral reef. It was formed off the northeast shore of Australia and is about 1,250 miles (2,011 km) long. It is so long that it can be seen from outer space.

A barrier reef is a coral formation in the ocean that grows parallel to a shoreline. It is separated from land by a deep lagoon. A barrier reef creates a barrier between the lagoon and the open sea. Over 2,000 of the coral formations, or colonies, that make up the Great Barrier Reef are barrier reefs. About 760 of the formations are fringing reefs. Fringing reefs grow closer to the shoreline. They have shallow bodies of water between them and the land. Some sections of the Great Barrier Reef are only 10 miles (16.1 km) from land. Some are as far as 100 miles (161 km) off the shore.

The Great Barrier Reef is home to many underwater plants and animals. In this book, we will explore this complex undersea world using graphs and charts to solve problems.

# What Is the Reef Made Of?

The Great Barrier Reef is made mostly of the hardened limestone skeletons of small sea creatures called coral polyps (PAH-lups). Coral polyps, which are related to jellyfish, live inside skeletons they create. The term "coral" is used when referring to both the creatures and their skeletons. There are about 2,500 different species of corals. Over 400 species can be found in the Great Barrier Reef. Corals can be as tiny as a pinhead or as large as 1 foot (0.3 m) in diameter. They are tube shaped and have saclike bodies. At one end of the tube is a mouth that is circled by stinging **tentacles**. Corals do not have backbones. They use matter from seawater to build hard, cup-shaped skeletons around themselves.

The stonelike structures that make up a reef are the hardened skeletons of dead corals. Living corals attach to these dead skeletons and sometimes even to each other, forming coral colonies.

Coral is permanent once it settles. Some coral in the Great Barrier Reef is thought to be 18 million years old! Corals grow best in warm, clear, shallow water that allows plenty of sunlight into the water. A reef grows very slowly, perhaps only $\frac{1}{2}$ inch (1.3 cm) each year. Corals extend out of their skeletons at night to find food. They use their tentacles to reach out and sting prey with poison. Small corals eat tiny water organisms called **plankton**. Larger corals can eat small fish. Over time, separate coral colonies grow together to form a reef. These colonies have many different shapes, sizes, and colors of coral.

White stony coral is a variety of coral found on the Great Barrier Reef.

Hard or stony coral, the kind of coral that reefs are made of, has no color of its own. A reef, though, can be many colors: tan, orange, blue, green, purple, red, white, and yellow. Where does the color come from? Coral has a special relationship with an organism called algae. This **symbiotic** relationship helps each of them. The algae get a place to live and grow, and in return they help feed the coral. The tentacles of corals often cannot catch enough food to sustain life. Algae grow inside corals and use **photosynthesis** to transform sunlight into sugar that feeds both the algae and the corals. The algae are what give coral its color.

Because algae need sunlight for photosynthesis, corals are usually found in shallow waters where the sunlight can reach them. They also prefer clear water. Corals may die when they expel the algae living in them due to disease or to changes in the temperature or salt content of the water.

Look at the scatter plot. It shows the growth of coral colonies in different depths of water. Which encourages coral growth of more than 10 millimeters: a depth less than 8 meters or more than 8 meters?

Locate 8 meters on the bottom of the chart. This will be our dividing line. Locate 10 millimeters on the left side of the chart. Compare the growth of coral colonies on each side of the dividing line. Coral colonies grew 10 millimeters or more at depths of less than 8 meters.

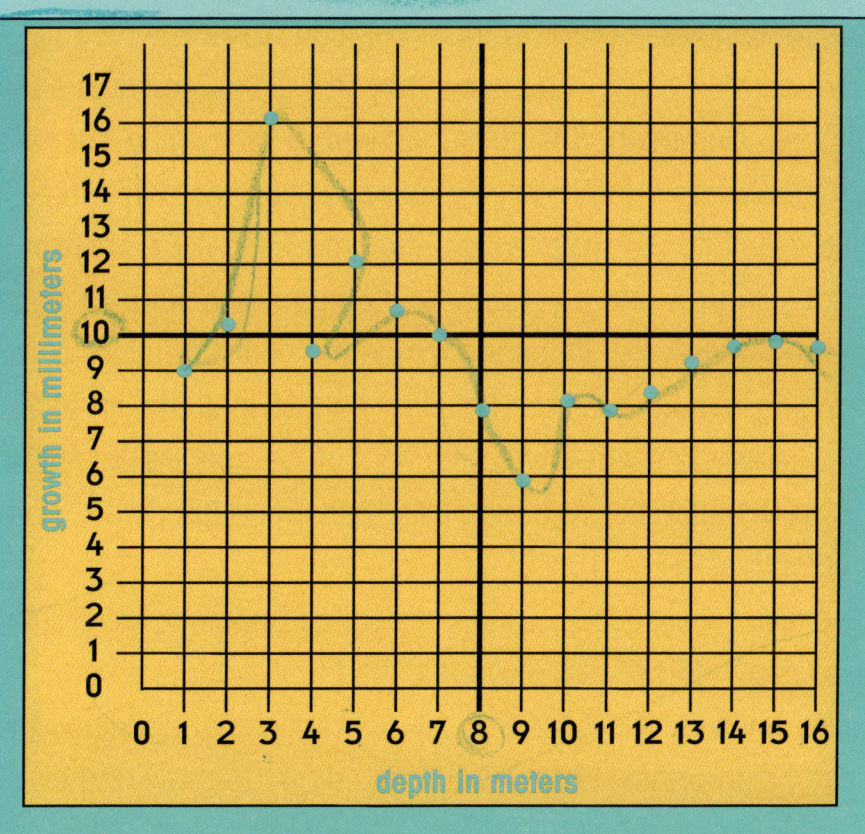

In addition to water depth, a number of different factors can affect the growth of coral colonies, including wave action, water temperature, water clarity, and natural enemies.

Corals reproduce at different times of the year on different reefs; for the Great Barrier Reef, it is in late spring and early summer. Since the Great Barrier Reef is south of the equator and its seasons are opposite those in the Northern Hemisphere, mating takes place in the months of November and December. Because hard corals do not move once they settle, they reproduce using 2 special methods. One method is called **spawning**. Spawning happens for just 1 or 2 nights following a full moon. Corals release male and female **reproductive** cells into the water, where they randomly join. **Larvae** are produced that will grow into corals under the right conditions. One colony of corals may produce several thousand larvae per year. The sight of milky white larvae swimming through the water is often said to look like an underwater snowstorm. The larvae swim toward the light at the surface. Usually between 2 days and 3 weeks later, they start to settle, become adult corals, and start a new coral colony.

Corals also reproduce by a process called budding. A coral divides and makes a copy of itself that stays attached to the original coral. The new coral matures, divides, and goes through the process of budding. This continuous addition of new coral allows the colony to grow and expand. Reproduction by budding helps to repair a reef when it has been damaged by storms and intense water movement.

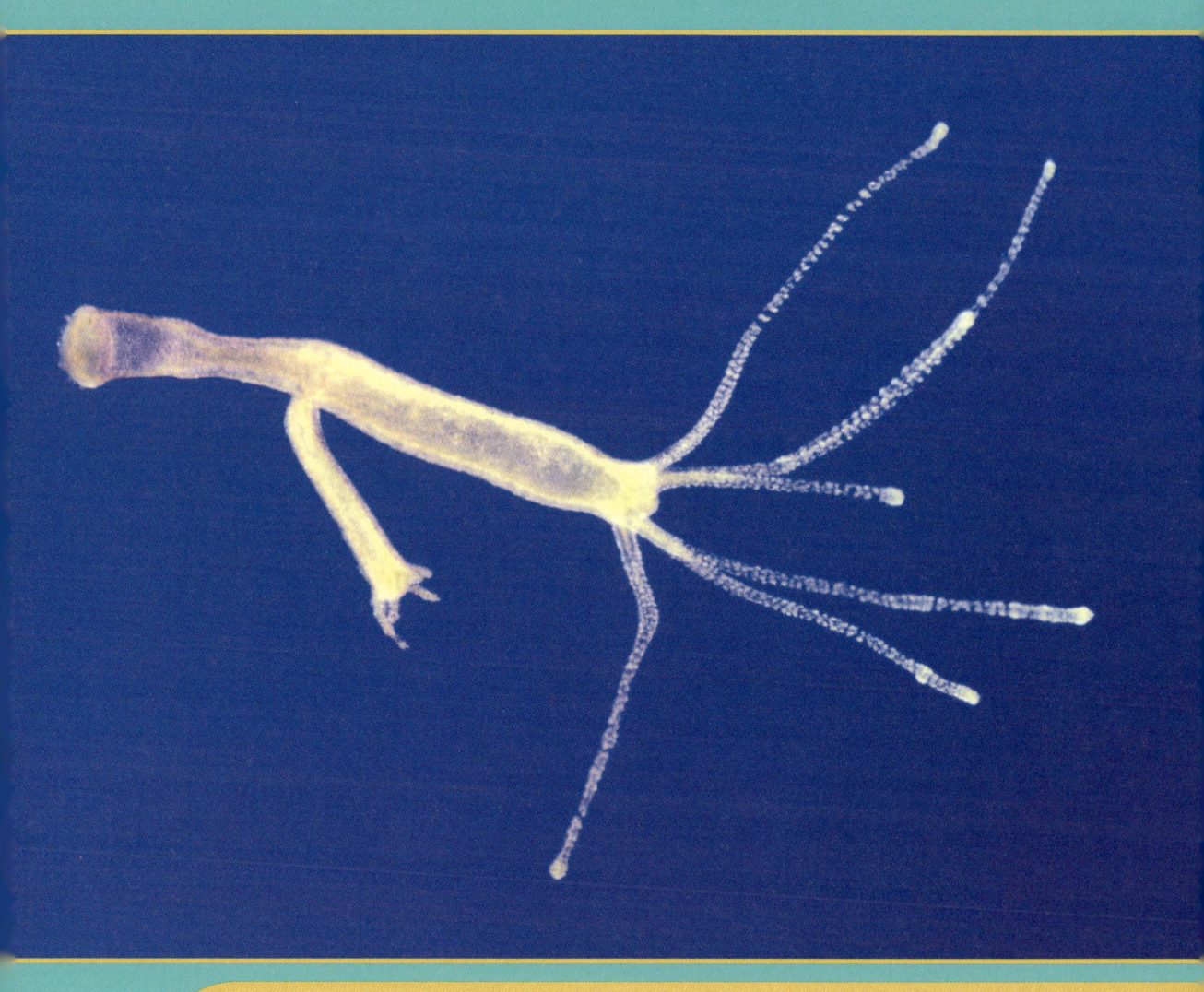

This coral larva will be lucky to survive its first days of life. Once larvae settle, though, their chances of survival are good. The time it takes for new corals to begin reproducing depends on the species. It can take up to 8 years.

# Creatures of the Reef

There are many interesting and unusual animals that live in and near the Great Barrier Reef. Over 1,500 species of fish and 5,000 **mollusk** species get their food from corals, algae, and the reef's plants such as sea grass and trees called mangroves. A very interesting mollusk, called a sea slug, can be found at the Great Barrier Reef. Sea slugs can be up to 12 inches (30 cm) long. They get their food by using poisons to catch their prey. They get the poisons from eating poisonous creatures such as some corals and jellyfish. They store the poisons from their food in their bodies until they are ready to use them. They also use the poisons to protect themselves from enemies.

The cone snail, or cone shell, is another mollusk of the Great Barrier Reef. The cone snail is very dangerous. It has a sharp stinger that gives off poison that kills the cone snail's prey, such as worms and others mollusks. A cone snail's poison is so deadly it could kill a person.

The reef is a large **ecosystem** in which each creature relies on other creatures for food and population control. Crabs, jellyfish, octopi, sea turtles, sea urchins, dolphins, and whales can also be found near the reef. Many kinds of birds—including pelicans, herons, and ospreys—feed on these sea creatures. You can see some of the diverse life of the reef on the chart on the opposite page.

If there are 7 species of sea turtles in the world, what percentage of all sea turtle species can be found in the area around the Great Barrier Reef? The chart shows us that the Great Barrier Reef has 6 species of sea turtles. To find what percent this is of all sea turtle species, we can create a proportion. A proportion is a comparison of equal ratios.

$x$ = percent of all sea turtle species found around the Great Barrier Reef

$$\frac{6}{7} = \frac{x}{100}$$

Cross multiply:
$7x = 600$

Divide both sides by 7:

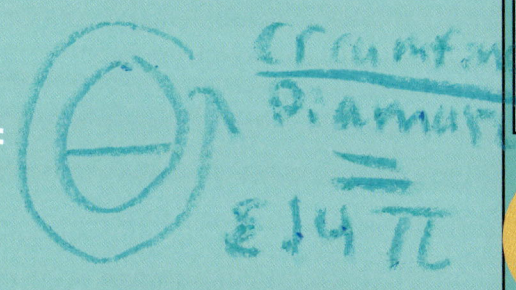

$$\frac{7x}{7} = \frac{600}{7}$$

$$x = \frac{600}{7}$$

$x = 85.71\%$

Round your answer to the nearest whole number: 86%.

Eighty-six percent of all sea turtle species can be found in the area of Australia's Great Barrier Reef.

Can you think of a different way this problem can be solved?

| Great Barrier Reef Estimated Species | |
|---|---|
| fish | 1,500 |
| mollusks | 5,000 |
| birds | 242 |
| hard corals | 400 |
| sea turtles | 6 |
| sea snakes | 14 |
| seaweed | 500 |
| sea grass | 15 |

Each of these animals and plants is part of the ecosystem of the Great Barrier Reef.

This green sea turtle is swimming in the waters near the Great Barrier Reef.

Just as coral and algae have a symbiotic relationship, so do other creatures of the Great Barrier Reef. **Sea anemones** and anemone fish are one example. Sea anemones, like corals, have stinging, poisonous tentacles. The poison of sea anemones has rarely caused serious harm to people. Sea anemones look like large coral polyps without skeletons. Sea anemones, unlike corals, can move very slowly by sliding. The anemone fish has a thick liquid coating that shields it from the anemone's poison. It lives among the sea anemone's tentacles. Living here protects it from predators, especially at night when the sea anemone closes up and keeps the fish inside it. In return, the anemone fish eats organisms harmful to the sea anemone and guards it from some of its enemies, like the butterfly fish.

Another interesting animal of the Great Barrier Reef is the blue-ringed octopus. The blue-ringed octopus changes color when it is disturbed or senses danger. This usually brown or yellow octopus is not very large. There are 4 species of blue-ringed octopi. The most common is about the size of a golf ball and weighs about 1 ounce (28 g). The others can grow to be 8 inches (20 cm) with their tentacles extended. When threatened, blue rings appear on the octopus's body. The electric-blue, glowing rings are a sign that the octopus is ready to protect itself. The blue-ringed octopus has 2 types of poison. One is less powerful and is used to capture prey, like crab and fish. The other is extremely deadly and used for protection. If bitten with the parrotlike beak of the blue-ringed octopus, a person can die in less than 10 minutes.

A blue-ringed octopus has enough poison to kill 26 adults!

# Endangered Species

The Great Barrier Reef is home to several **endangered** species. Among these are green sea turtles, which are often found close to shore in lagoons near the Great Barrier Reef. These animals have been hunted, their nesting beaches have been taken over by human activity, and many of them have been captured, harmed, or killed in fishing nets.

Turtles are reptiles, which means they are **cold-blooded**, breathe air, and have scales. Green sea turtles are light to dark brown in color. Why, then, are they called "green"? They get their name from the color of their fat. It has been turned green by the sea grasses and algae they eat!

Green sea turtles are among the largest sea turtles. They can weigh up to 400 pounds (182 kg) and be 36 to 43 inches (91 to 109 cm) long. They have a shell, but unlike land turtles, they have flippers instead of feet. They can swim up to 35 miles (56 km) per hour. When green sea turtles need to get rid of excess salt they have taken in from seawater, they shed tears. They have special organs behind their eyes that allow them to "cry" out the salt.

A female green sea turtle comes ashore to lay her eggs. She digs a hole in the sand with her front flippers, then lays 100 to 120 eggs. She covers all the eggs with sand and returns to the sea. About 2 months later, the young turtles, or **hatchlings**, hatch and dig out of the sand. They then make their way to the water. Hatchlings risk being captured and eaten by crabs, birds, or other predators. Many never make it to the sea.

The sex of a green sea turtle is determined by the temperature of the sand in which the egg is laid. The chart below shows the temperature of several areas of sand and the sex of the green sea turtles that hatched in those areas. Based on the information on the chart, which of the following statements may be true?

**Green Sea Turtle Sex and Nest Temperature**

| temperature | sex |
|---|---|
| 86°F (30°C) | female |
| 82°F (28°C) | female |
| 79°F (26°C) | male |
| 77°F (25°C) | male |
| 75°F (24°C) | male |

a. Most female green sea turtles are formed in sand with temperatures greater than 81°F (27°C).

b. Most male green sea turtles are formed in sand with temperatures greater than 81°F (27°C).

According to the information on the chart, female green sea turtles were hatched from eggs laid in sand with a temperature greater than 81°F (27°C). Therefore, statement *a* may be true.

This green sea turtle hatchling is coming out of its egg in the sand.

The dugong of the Great Barrier Reef is also an endangered species. Dugongs are unique marine mammals. They can live for 70 years or more. Dugongs can be 10 feet (3 m) long and can weigh over 800 pounds (363 kg). This large animal is sometimes called the sea cow because it grazes on sea grasses and eats mostly plants, just like cows on land graze on grasses and eat mostly plants. Dugongs have round heads, small eyes on the sides of their heads, and large snouts. They have very poor eyesight but very good hearing. Special sensitive bristles in their mouths allow them to feel and grasp the grasses as they swim past. Since they must breathe air, dugongs can only stay underwater for a few minutes at a time.

Most of the world's dugongs live near Australia, with its warm waters and plentiful plant life. It is estimated that there are only a few thousand dugongs left around the Great Barrier Reef. Researchers believe that a large number of dugongs have been killed by passing boats, fishing nets, pollution, and **poachers** because of the increasing human population and activity along the coast. It is difficult to increase the population of dugongs because they need many years to mature before they can give birth. Also, dugong mothers give birth only once every $2\frac{1}{2}$ to 5 years and have only 1 baby at a time.

*The dugong, shown here, is most closely related to another creature of the sea: the manatee. The land animal a dugong is most closely related to is the elephant.*

| Dugong Reproduction | |
|---|---|
| age at which females can mate | 6–17 years |
| age at which males can mate | 4–16 years |
| amount of time baby is inside mother | 13–15 months |
| number of babies | 1 |
| maximum rate of population increase per year | 5% |

Use the information from the chart to solve this word problem. The population of dugongs around the Great Barrier Reef is about 4,000. If the maximum rate of population increase occurred in 1 year, what would be the total population of dugongs at the end of that year? To solve this problem, find the maximum rate of population increase possible on the chart.

The maximum rate of increase is 5%.

Change 5% into a decimal:

5 ÷ 100 = .05

Multiply 4,000 by .05:   
```
4,000 dugongs
x   .05
200.00 dugongs
```

This is the maximum increase in the number of dugongs in 1 year.

Add the maximum increase to the 4,000 original dugongs:
```
  4,000
+   200
  4,200
```

The total population of dugongs at the end of that year would be 4,200, assuming that no dugongs died during the year.

# Enemies of the Great Barrier Reef

The Great Barrier Reef is home to many species of underwater life. Without the reef, many of these species would not survive. Scientists have discovered some threats to the life of the Great Barrier Reef. One of the most common and serious natural threats in recent years is the population explosion of crown-of-thorns starfish. These starfish of the Great Barrier Reef are usually brown or reddish gray in color with red-tipped spines. The spines of the crown-of-thorns starfish are very sharp.

These creatures are a natural part of the reef's ecosystem and eat reef-building corals. In great numbers, crown-of-thorns starfish can cover and easily kill a reef. The increase in the starfish population in recent years has caused a change in the balance of life on the reef. Large areas of living corals have been destroyed.

Scientists are not sure why there are now so many crown-of-thorns starfish. There are several possible explanations. A natural cycle may be occurring—starfish increase in population and after a period of time will decrease. The fish and mollusks that feed on crown-of-thorns starfish may have been overfished by humans. Human activity on the coast of Australia may have led to the release of **nutrients** into the coastal waters. These nutrients may have increased the amount of plankton. This would enable greater numbers of crown-of-thorns starfish larvae to survive and mature. They would then feed on the reef.

**Coral Reef Recovery**

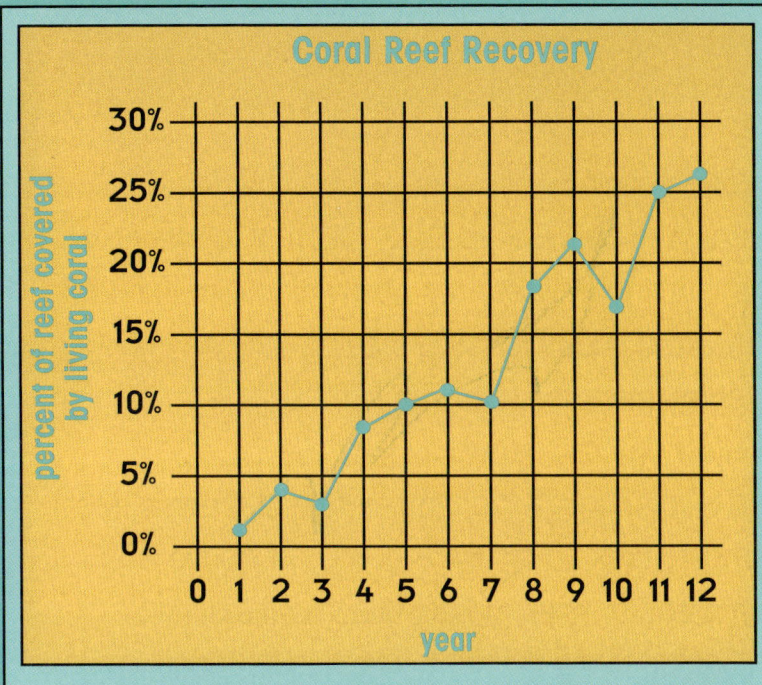

This line graph shows coral reef recovery for 12 years following an attack by crown-of-thorns starfish. Use the information on the graph to solve the following word problem:

On a healthy reef, living corals usually cover 20% to 40% or more of the reef surface. Crown-of-thorns starfish can reduce that number to less than 1%. According to the graph, how long did it take this coral reef to recover to a normal percentage of living corals after an attack by crown-of-thorns starfish?

A normal percentage of living corals would be 20% or more. Locate this number on the left side of the graph. Then find the year at which the colony reached at least 20% by looking at the years across the bottom of the graph. The coral reef recovered a normal percent of living corals after about 9 years.

Crown-of-thorns starfish are among the few sea creatures that eat living corals.

People are also enemies of the Great Barrier Reef. Much of the lands on Australia's coasts were once wetlands that filtered the freshwater from the land before allowing it into the ocean. Now, many farms are being built on the coasts. These farms use fertilizer to enrich the soil. The fertilized soil seeps into the ocean and has a negative effect on coral growth.

Another problem for coral is changes in the temperature of the sea. Most corals do not thrive in water temperatures below 64°F (18°C) or above 85°F (30°C). If the water temperature drops or rises, corals become "bleached." This means the coral rejects the algae living inside it, which is its most valuable source for food. Without the algae, corals become a whitish color. Bleached corals will die if the water temperature does not return to normal.

The temperatures around the Great Barrier Reef do not usually vary much throughout the year. However, in recent years, the temperatures have been rising due to **global warming**. People using fuels like gasoline, oil, and coal, which add heat-trapping gases to the air, make global warming worse. The destruction of plant life, such as that found in rain forests, removes a natural way of cleaning the air. The heat-trapping gases in the air have helped to change weather patterns and temperatures worldwide.

The pie charts show what percents of coral reefs in the Great Barrier Reef suffered high, moderate, and low levels of bleaching as of 2002. Based on the information on the charts, which type of reef had the greatest area with a high level of bleached coral, inshore or offshore? Calculate the percent difference. Then find the percent differences for the areas of the inshore and offshore reefs that suffered moderate and low levels of bleaching. Look at and compare the percent differences for all 3 levels.

**Inshore Fringing Reefs**

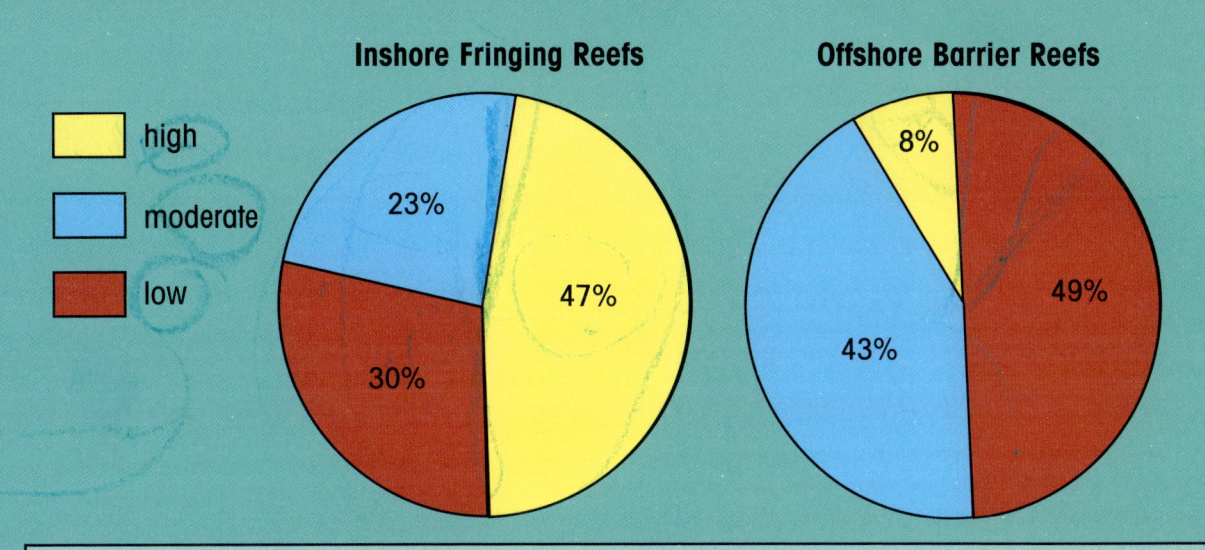

- high (yellow)
- moderate (blue)
- low (red)

Inshore Fringing Reefs: 47% high, 23% moderate, 30% low

**Offshore Barrier Reefs**: 8% high, 43% moderate, 49% low

We see that 47% of the fringing or inshore reefs had a high level of bleached coral. Eight percent of the barrier or offshore reefs had a high level of bleached coral.

Therefore, a greater portion of the inshore reefs had a high level of bleached coral. To find out the percent difference, subtract 8% from 47%.

$$\begin{array}{r} 47\% \\ -\phantom{0}8\% \\ \hline 39\% \end{array}$$

Thirty-nine percent more of the inshore reefs suffered a high level of bleaching.

Now, compare the percent of the inshore and offshore reefs showing moderate and low levels of bleached coral and calculate the percent difference.

**Moderate level:**
Inshore: 23%
Offshore: 43%
Subtract 23% from 43%.

$$\begin{array}{r} 43\% \\ -23\% \\ \hline 20\% \end{array}$$

Twenty percent more of the offshore reefs suffered a moderate level of bleaching.

**Low level:**
Inshore: 30%
Offshore: 49%
Subtract 30% from 49%.

$$\begin{array}{r} 49\% \\ -30\% \\ \hline 19\% \end{array}$$

Nineteen percent more of the offshore reefs had a low level of bleaching. When we look at the percent differences for all 3 levels, the smallest difference between inshore and offshore reefs was at the low level of bleaching.

# Shipwreck!

In 1789, the famous English ship HMS *Bounty* traveled the waters of the South Pacific. Its crew had set out to find, learn more about, and obtain **breadfruit** plants. The sea voyage was very difficult due to unfavorable weather and disagreements among the crew. The ship's captain, William Bligh, found himself in a dangerous situation when most of his crew mutinied. They forced Bligh and about 18 crew members who were loyal to him into a small boat. The mutineers then sailed away in the *Bounty*. Bligh returned to England about 1 year later and told the king what had happened. The king wanted his ship back and the crew punished. He sent Captain Edward Edwards and the HMS *Pandora* to fulfill this mission.

The HMS *Pandora* set sail in 1790. The crew found part of the *Bounty* crew on the island of Tahiti. After capturing them, Edwards had a wooden box made for his prisoners on the deck of the ship. The *Pandora* sailed the South Pacific looking for the *Bounty* and the rest of her crew. On August 29, 1791, the *Pandora* collided with a coral reef while trying to sail through the Great Barrier Reef. The ship was damaged beyond repair and started to take on water. The prisoners in the box cried to be freed. Finally, orders were given to release them. Thirty-five men, including 4 prisoners, died when the ship finally sank.

The *Pandora* rested on the bottom of the ocean and was gradually buried by layers of sand. It lay untouched for over 100 years.

This reef shark swims around a sunken ship.

25

In 1977, a team of researchers—which included American and Australian filmmakers—and the Royal Australian Air Force searched for the *Pandora*. The researchers tried to figure out the route the *Pandora* might have taken through the reef. The sea was searched with powerful metal detectors from the Royal Australian Air Force. Divers were sent down to investigate. On November 16, 1977, the team found the ship. The most exciting part of the discovery was that much of the ship had been preserved because it had been covered with sand. The *Pandora*'s anchor, several canons, and objects from inside the frame of the ship were covered with coral.

Thousands of items were found that give us a glimpse into sailors' lives during the 1700s. Many items were found in cabins located on the lower decks of the ship. These items included clothing, furniture, jewelry, tableware, souvenirs from islands that had been visited, tools, jars of spices, and even skeletal remains of crew members. In the captain's storeroom, the researchers found corked bottles that had originally held "spruce essence," a liquid made from the needles and branches of spruce trees. "Spruce essence" was used in the eighteenth century to make a drink that would fight off a disease called **scurvy**, which often affected sailors.

**Cost of Expeditions**

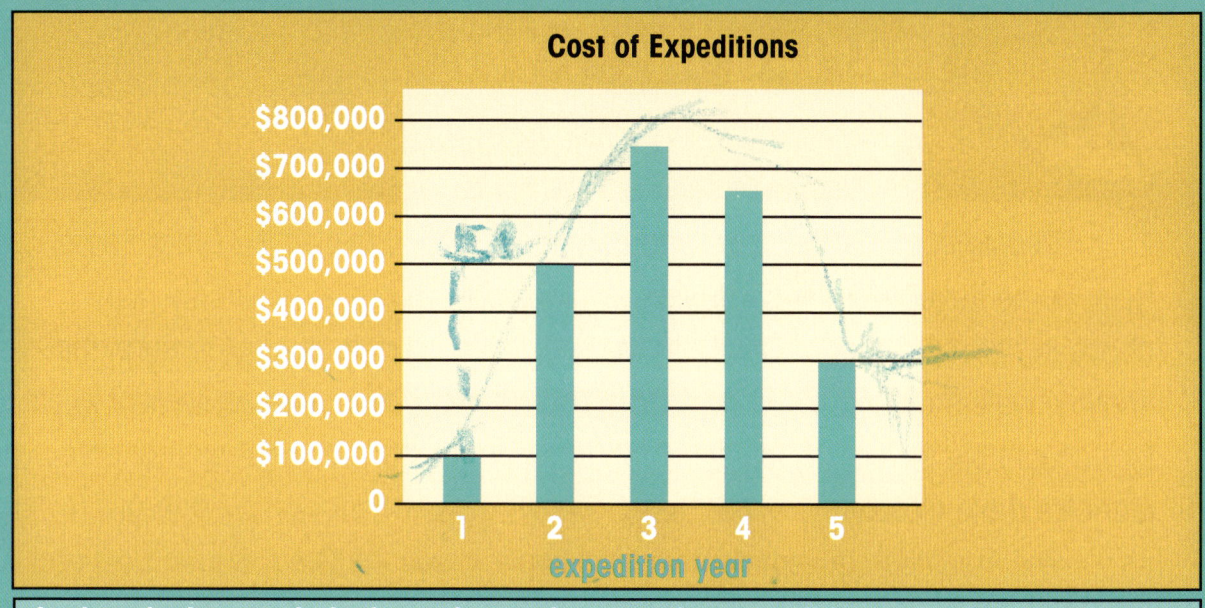

(Bar graph showing costs by expedition year:
- Year 1: $100,000
- Year 2: $500,000
- Year 3: $750,000
- Year 4: $650,000
- Year 5: $300,000)

Look at the bar graph. It shows the yearly costs of an expedition to the *Pandora*. The amounts depend on the equipment and supplies needed, the number of crew members involved, and other variables. Let's say a research team received $3 million for their expeditions for 5 years. Would that be enough to cover the cost of the expeditions? How much money would be left over or how much more money would be needed to complete 5 years of expeditions?

First, use the information on the bar graph to find the total cost for 5 years of expeditions.

$$\begin{array}{r} \$100,000 \\ 500,000 \\ 750,000 \\ 650,000 \\ +\ 300,000 \\ \hline \end{array}$$

$2,300,000 is the cost of expeditions for 5 years.

Since this sum is less than $3,000,000, there would be money left over. To find out how much money was left, subtract the cost for 5 years of expeditions from the original $3,000,000.

$$\begin{array}{r} \$3,000,000 \\ -\ 2,300,000 \\ \hline \$\ \ \ 700,000 \end{array}$$

There would be $700,000 remaining.

**27**

# Touring the Reef

People from all over the world come to Australia to visit the Great Barrier Reef. Some travel over the coral reefs in glass-bottom boats. Others fly high above in helicopters for a view from the sky. Some tourists snorkel through the shallow waters around the reefs. Others use air tanks and other equipment to go far beneath the surface, where they can observe the reef's colorful animal and plant life more closely.

The Australian government wants people to see the Great Barrier Reef and appreciate its importance. It also wants people to be aware of the Great Barrier Reef's delicate ecosystem. Australians watch carefully to be sure that the reefs are protected from excess fuel in the water or danger to the wildlife. Only certain businesses are given permission to be part of the tourism trade that takes people out to the reefs. The guides are specially trained so that the underwater life is disturbed as little as possible.

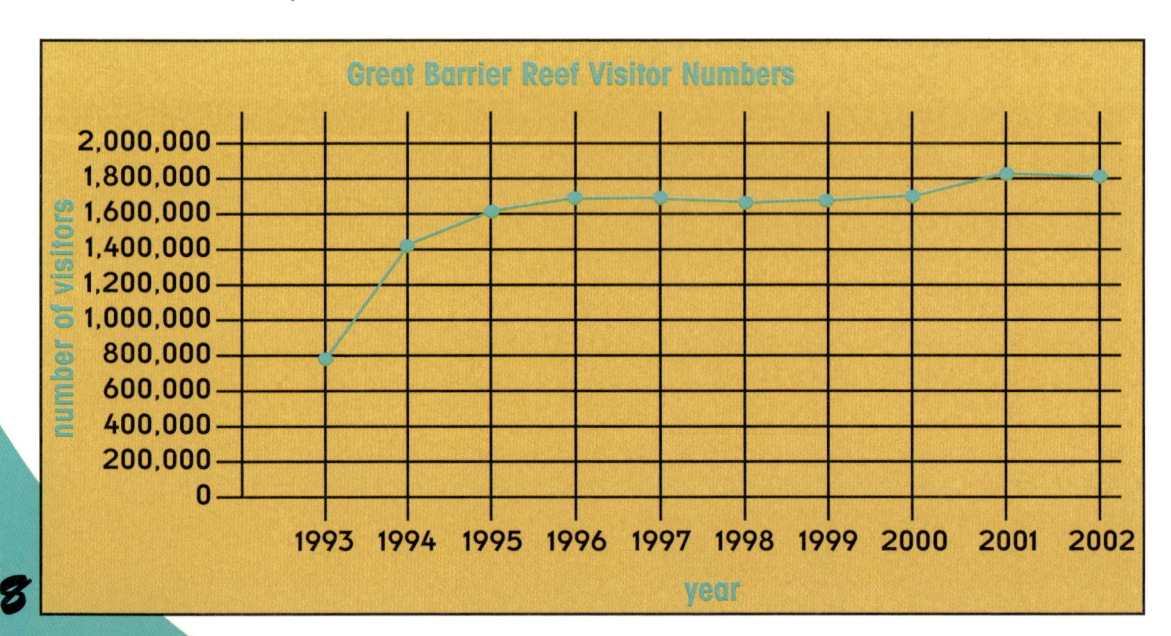

**Great Barrier Reef Visitor Numbers**

Look at the line graph on the opposite page. It shows the number of visitors to the Great Barrier Reef from 1993 through 2002. Use the information from the graph to solve this word problem.

What is the increase in the number of visitors from 1993 to 2002? What is the percent increase?

To solve this problem, determine the number of visitors for 1993 and 2002. Subtract to find the increase.

$$\begin{array}{r} 1{,}800{,}000 \text{ visitors in 2002} \\ -\ 790{,}000 \text{ visitors in 1993} \\ \hline 1{,}010{,}000 \text{ increase in visitors} \end{array}$$

The difference is 1,010,000 visitors.

To find the percent increase, write a fraction by putting the increase in visitors over the number of visitors in 1993. Then divide the numerator by the denominator. Simplify first by crossing out the zeros.

$$\frac{1{,}010{,}000}{790{,}000}$$

$$\begin{array}{r} 1.278 \\ 79\overline{)101.000} \\ -\ 79\phantom{.000} \\ \hline 220\phantom{00} \\ -\ 158\phantom{00} \\ \hline 620\phantom{0} \\ -\ 553\phantom{0} \\ \hline 670 \\ -\ 632 \\ \hline 38 \end{array}$$

To change the decimal into a percent, multiply by 100.

$$\begin{array}{r} 1.278 \\ \times\ 100 \\ \hline 127.800 \end{array}$$

There was a 127.8% increase in visitors.
Rounded to the nearest whole number, the increase in visitors was 128%.

## Valued by the World

In 1975, the Great Barrier Reef was declared a national park of Australia—the Great Barrier Reef Marine Park. Activities within the park are closely monitored, whether it is boating, diving, fishing, or research. In 1981, the Great Barrier Reef was named a World Heritage Area in recognition of its importance to Earth.

It is estimated that the reefs of the world provide food for about $\frac{1}{4}$ of all ocean life. If the Great Barrier Reef disappeared, we cannot be sure what the extent of the damage to Earth would be. The current situation involving crown-of-thorns starfish teaches us that a change in 1 population can greatly affect others. We cannot know how a permanent decrease in the population of hard corals might affect the human population. Some species of sea creatures and sea plants would cease to exist. We do know that what we do and how we treat Earth and its life-forms can have an impact on the natural resources and lifestyles we now enjoy.

Compiling information and organizing data on graphs and charts are ways scientists monitor and evaluate nature. Studying and gathering information about ocean waters, plant and animal life, and human influence can help us find ways to protect the Great Barrier Reef of Australia—a natural wonder of the world.

# Glossary

**breadfruit** (BRED-froot)  A tropical fruit that is similar to bread when baked.
**cold-blooded** (KOHLD–BLUH-duhd)  Having a body temperature that changes with the temperature of the surrounding environment.
**ecosytem** (EE-koh-sihs-tuhm)  The living and nonliving things that make up an environment and affect each other.
**endangered** (ihn-DAYN-juhrd)  In danger of becoming extinct.
**global warming** (GLOH-buhl WOHR-ming)  An increase in Earth's temperatures as a result of pollution.
**hatchling** (HACH-ling)  An animal that has recently come out of an egg.
**larva** (LAHR-vuh)  A young form of an animal.
**mollusk** (MAH-lusk)  A soft-bodied animal that usually has a hard shell.
**nutrient** (NOO-tree-uhnt)  Something needed for life and growth.
**photosynthesis** (foh-toh-SIHN-thuh-suhs)  The process carried out in plants of using sunlight, carbon dioxide, and water to make food.
**plankton** (PLANK-tuhn)  Tiny plants and animals that float in water.
**poacher** (POH-chuhr)  A person who captures or kills wild animals illegally.
**reproductive** (ree-pruh-DUHK-tihv)  Able to reproduce by joining together.
**scurvy** (SKUHR-vee)  A disease caused by a lack of vitamin C.
**sea anemone** (SEE uh-NEH-muh-nee)  A sea animal with tentacles. Sea anemones can be various colors.
**spawn** (SPAWN)  To release large numbers of male and female reproductive cells into the water.
**symbiotic** (sihm-bee-AH-tik)  A relationship between 2 organisms from which they both benefit.
**tentacle** (TEHN-tih-kuhl)  An armlike body part on an ocean animal. Tentacles are used to catch prey and often have poisonous stingers.

# Index

**A**
algae, 8, 12, 14, 16, 22
anemone fish, 14

**B**
barrier reef(s), 5, 23
bleached, 22, 23
bleaching, 23
blue-ringed octopus, 14
*Bounty*, 24
budding, 10

**C**
colony(ies), 5, 6, 9, 10
crown-of-thorns starfish, 20, 21, 30

**D**
dugong(s), 18, 19

**E**
ecosystem, 12, 20, 28
endangered, 16, 18

**F**
fringing reefs, 5, 23

**G**
global warming, 22
Great Barrier Reef Marine Park, 30
green sea turtle(s), 16, 17

**L**
lagoon(s), 5, 16
larvae, 10, 20

**P**
*Pandora*, 24, 26, 27
photosynthesis, 8
plankton, 6, 20
poison(s), 6, 12, 14
polyps, 6

**S**
sea anemone(s), 14
sea turtle(s), 12, 13, 16
spawning, 10
symbiotic, 8, 14

**T**
tentacles, 6, 8, 14

**W**
wonder(s), 5, 30